Contents

1 Introduction to Biosensors

A biosensor is fundamentally a device that combines a biological recognition element with a transducer. This biological element, which could be an enzyme, antibody, or even a whole cell, selectively interacts with a target analyte—the substance you're trying to detect, be it a bacteria, a toxin, or a specific molecule indicating food quality. The interaction triggers a measurable change in the transducer, a component that converts this biological signal into an electrical, optical, or other quantifiable signal. Think of it like a highly specific lock and key mechanism, where the biological element is the lock and the target analyte is the key, and the transducer is the mechanism that tells you if the key fits. This process allows for the sensitive and selective detection of various substances even in complex matrices like food products.

Electrochemical biosensors are a common type, employing electrodes to measure changes in current, voltage, or impedance caused by the biorecognition event. These are often relatively inexpensive and portable, making them suitable for on-site testing. Optical biosensors, on the other hand, rely on changes in light properties, such as absorbance, fluorescence, or reflectance, to signal the presence of the analyte. They often offer high sensitivity and can be used for multiplexed detection of several analytes simultaneously. Piezoelectric biosensors utilize the piezoelectric effect, where a change in mass on a crystal surface alters its vibrational frequency. This can be exceptionally useful for detecting small changes in mass associated with biological interactions. Each type presents its own advantages and limitations, the selection depending on the specific application and the nature of the target analyte. The choice is not arbitrary; factors like sensitivity, selectivity, cost, and portability all play a significant role.

Many biosensors operate based on affinity interactions, where the biological recognition element specifically binds to the target molecule. This is exploited in immunoassays, where antibodies are used to detect specific antigens, which can be bacterial toxins or pathogens. However, biosensors can also work on the basis of catalytic reactions. For example, enzymatic biosensors use enzymes to catalyze a specific reaction, producing a measurable signal directly proportional to the analyte concentration. Understanding the underlying principle of a particular biosensor is crucial for its appropriate application and accurate interpretation of results. Proper understanding allows one to effectively leverage the technology for precise and reliable analyses. A thorough comprehension of this is vital for the effective implementation of this crucial technology in the food industry.

A crucial aspect of biosensor design involves the immobilization of the biological recognition element onto the transducer surface. The method of immobilization must ensure that the bioelement retains its activity and specificity, while also allowing for efficient interaction with the target analyte. Common techniques include physical adsorption, covalent bonding, and entrapment within a matrix. The choice of immobilization method significantly impacts the sensor's performance and its longevity. A poorly immobilized element can lead to a decrease in sensitivity or selectivity, rendering the sensor ineffective. Therefore, careful consideration of immobilization strategies is critical for developing high-performance biosensors.

Furthermore, signal processing and data analysis are critical aspects of biosensor technology. Raw signals from the transducer must be processed to obtain accurate and reliable measurements. This often involves signal amplification, filtering, and calibration procedures. Appropriate data analysis techniques are necessary to extract meaningful information from the sensor output, such as analyte concentration and specificity. Advanced signal processing algorithms are

continually being developed to improve the accuracy, sensitivity, and speed of biosensor measurements. A well-designed system integrates these elements to deliver reliable results.
Ultimately, biosensor technology offers a powerful set of tools for addressing various food safety and quality challenges. Their inherent sensitivity, speed, and selectivity makes them particularly well-suited for detecting low levels of contaminants, ensuring quick results, and simplifying quality control procedures. The ability to perform rapid, on-site testing is particularly crucial in preventing outbreaks of foodborne illnesses. The adoption and integration of biosensor technology into routine food safety protocols have the potential to dramatically improve public health and provide a high level of confidence in the safety of the food supply. Continuous improvement and innovations in this field will further enhance its already substantial impact. The future of food safety is inextricably linked to the continued advancement and wider adoption of biosensor technologies.

1.1 Biosensor Principles

A biosensor, at its core, is a device that combines a biological recognition element with a transducer. The biological element, often an enzyme, antibody, or nucleic acid, specifically interacts with the target analyte—the substance we're trying to detect in food, for instance. This interaction generates a signal that's then converted into a measurable output by the transducer. This could be a change in electrical current, light emission, or a mechanical response, depending on the type of biosensor. The key is the specificity and sensitivity of this interaction, crucial for accurate and reliable detection. Think of it like a highly specialized lock and key mechanism, where only the target molecule "fits" and triggers the signal.
Understanding this fundamental principle is key to appreciating the vast potential of biosensors in food safety. Imagine a quick, easy test that can detect harmful bacteria in a sample of milk within minutes, eliminating the need for lengthy laboratory analyses. That's the power of biosensor technology. It offers rapid, accurate, and often

portable solutions to challenges that traditionally required extensive time and resources. This translates to immediate benefits for both producers and consumers.

The transducer is the other crucial component. This component transforms the biological interaction into a quantifiable signal. Different types of transducers cater to different needs and offer various advantages. Electrochemical transducers, for example, measure changes in electrical current or potential, making them suitable for detecting substances that induce redox reactions. Optical transducers, on the other hand, measure changes in light absorption, reflection, or fluorescence, often offering high sensitivity. Piezoelectric transducers, relying on changes in mechanical properties, have found applications in detecting mass changes associated with the binding of target molecules. Choosing the right transducer depends greatly on the specific application and the nature of the analyte being detected.

The sensitivity of a biosensor refers to its ability to detect very low concentrations of the target analyte. High sensitivity is vital, especially when detecting trace amounts of harmful substances in food. Specificity, on the other hand, indicates the biosensor's ability to selectively identify the target analyte without interference from other substances. This is critical because food matrices are complex mixtures containing a vast array of compounds. A highly specific biosensor won't be misled by other molecules present in the sample, providing a clear and accurate result.

The design and optimization of biosensors require careful consideration of numerous factors. The choice of the biological recognition element directly affects specificity and affinity towards the target. The immobilization method used to attach the biological recognition element to the transducer surface must be stable and ensure optimal accessibility for the target analyte. Furthermore, the signal transduction process needs to be optimized for maximum sensitivity and minimal background noise, often requiring calibration and careful handling of the instrument and sample. A poorly designed or poorly calibrated biosensor could produce unreliable data.

Ultimately, biosensor technology's success rests on its ability to translate biological recognition into a practical, measurable signal. Understanding the interplay between the biological recognition element and the transducer is crucial for developing effective biosensors for diverse applications. This understanding, coupled with careful optimization of the biosensor design and operating conditions, will determine the sensitivity, specificity, and overall reliability of the device. And this accuracy is paramount in ensuring food safety and maintaining consumer confidence in the food supply chain.

Furthermore, the future of biosensor technology looks bright. Advancements in nanotechnology are leading to the development of smaller, more sensitive, and more portable devices. Miniaturization allows for faster analysis and more widespread applications. Integration with microfluidics and other micro-scale technologies creates systems that are user-friendly and suitable for on-site testing. Such portable and user-friendly biosensors are crucial for monitoring food quality in remote areas or during transportation, thus providing real-time information and quick response to potential hazards.

However, challenges remain. Cost-effectiveness and standardization are crucial for widespread adoption of biosensor technology. Regulatory approval processes can be complex and time-consuming, creating hurdles for the commercialization of new devices. Ensuring the longevity and stability of the biological recognition element is another hurdle that scientists are continuously tackling. Despite these challenges, continued research and development promise to overcome these obstacles, paving the way for a future where biosensors play an even more significant role in food safety and quality control. The integration of advanced data analysis and machine learning techniques further enhances the capabilities of biosensors, enabling predictive modelling and improving the interpretation of results. These developments suggest a promising future for this technology in transforming the food industry.

1.2 Types of Biosensors

Electrochemical biosensors represent a cornerstone of modern food safety analysis. They leverage the principles of electrochemistry to detect specific biomolecules, translating biological interactions into measurable electrical signals. These signals, often changes in current or voltage, are directly proportional to the concentration of the target analyte, be it a bacterial toxin, a pesticide residue, or a spoilage indicator. The simplicity and cost-effectiveness of electrochemical platforms, coupled with their portability and potential for miniaturization, make them highly attractive for on-site food quality assessment. Think of rapid screening for E. coli contamination in a processing plant, or quick checks for aflatoxins in harvested crops; electrochemical biosensors are ideally suited for these scenarios. Their sensitivity, however, is often limited, requiring careful optimization and calibration to ensure reliable results, and the effects of interfering substances in complex food matrices can pose challenges to their accuracy.

Optical biosensors offer a powerful alternative, relying on changes in light properties to signal the presence of target molecules. These changes can manifest as alterations in absorbance, fluorescence, or refractive index, providing a diverse range of detection modalities. For instance, surface plasmon resonance (SPR) biosensors detect binding events through changes in the refractive index near a sensor surface, offering label-free and real-time monitoring capabilities. Optical fibers can be integrated into sophisticated sensor arrays, creating highly sensitive and multiplexed platforms capable of simultaneously detecting various analytes. The high sensitivity and selectivity of many optical systems make them particularly useful in detecting trace amounts of toxins or pathogens. While their sensitivity is often superior to electrochemical methods, they can be more complex to operate and require specialized equipment, increasing both the cost and expertise required for effective implementation.

Piezoelectric biosensors operate on a different principle entirely: they measure changes in the resonant frequency of a piezoelectric crystal as a result of mass changes on its surface. When a biomolecule binds to the crystal's surface, the added mass causes a shift in its resonant frequency, directly proportional to the mass of the bound molecule. This approach offers a label-free and highly sensitive method for detection, suitable for a variety of analytes. They offer exceptional sensitivity and are often used in detecting very low concentrations of foodborne pathogens or toxins. However, piezoelectric biosensors are more susceptible to interference from non-specific binding events and environmental factors, demanding careful design and control of the sensing environment to achieve reliable and accurate results. This makes them perhaps less ideal for rapid, on-site analysis in highly variable settings.

Thermal biosensors operate on the principle of heat generation or absorption accompanying a specific biochemical reaction. These devices monitor minute temperature fluctuations to detect the presence of the target analyte. Microcalorimetry, a powerful technique, is frequently employed to analyze metabolic activity in microorganisms, indicating the presence of pathogens in food samples. The inherent advantages of using these systems are their non-destructive nature and high sensitivity for certain applications. However, interpreting the thermal signals can be complex and requires sophisticated data analysis, and the overall detection time may be longer compared to other methods.

Beyond these key types, several hybrid systems combine features from various biosensor categories, aiming to maximize their benefits and overcome limitations. For example, integrating electrochemical detection with optical signal enhancement can result in improved sensitivity and broader applicability. The ongoing advancements in nanotechnology and material science constantly lead to the emergence of novel sensing platforms, pushing the boundaries of what's achievable in terms of sensitivity, specificity, portability, and cost-effectiveness. This continual evolution ensures that biosensors will continue to play an increasingly vital role in advancing food safety and quality control, leading to safer and more nutritious food for everyone.

2 Biosensors in Foodborne Pathogen Detection

Rapid and sensitive detection of foodborne pathogens is paramount for ensuring food safety and preventing outbreaks of foodborne illnesses. Traditional methods, often slow and laborious, are being replaced by biosensors, offering faster, more efficient alternatives. Electrochemical biosensors, for example, leverage the change in electrical current or potential as a biological recognition element interacts with a target pathogen. This change is then measured and correlated to the pathogen concentration, providing a quantitative result. The simplicity and cost-effectiveness of electrochemical methods make them attractive for widespread adoption in food safety testing. Furthermore, advancements in nanotechnology have led to the development of highly sensitive electrochemical biosensors, capable of detecting pathogens at very low concentrations. This increases the sensitivity of the assay, reducing the likelihood of false negatives.

Optical biosensors utilize changes in light properties, like absorbance, fluorescence, or refractive index, to detect pathogens. Surface plasmon resonance (SPR) biosensors, for example, measure changes in the refractive index near a gold surface upon binding of a target pathogen to a capture antibody immobilized on that surface. These changes are incredibly sensitive and can detect even minute amounts of pathogens. This technology is suitable for detecting a wide array of bacterial and viral contaminants, offering high specificity. The relatively simple operation and high throughput potential of SPR make it a significant advancement in pathogen detection. However, the initial investment in instrumentation is substantial compared to other techniques.

Immunological methods integrated into biosensor platforms, such as ELISA (Enzyme-Linked Immunosorbent Assay), offer high specificity and sensitivity. Miniaturized ELISA-based biosensors are being developed for portable, point-of-care applications. These reduce the need for extensive laboratory equipment and trained personnel, making on-site testing feasible. This allows rapid testing in diverse locations, from farms and processing plants to retail outlets. However, the reliance on antibodies requires careful selection and optimization of these antibodies for each target pathogen to achieve optimal sensitivity and avoid cross-reactivity. The stability and shelf life of the reagents are also critical considerations for reliable performance.

Polymerase Chain Reaction (PCR)-based biosensors offer incredible sensitivity and speed for detecting pathogens even in complex food matrices. Real-time PCR allows for direct monitoring of the amplification process, providing quantitative data on the number of target pathogen DNA copies. Integrating PCR with microfluidic devices allows for miniaturization and automation, further enhancing speed and efficiency. The ability to identify specific genetic sequences allows for differentiation among closely related strains, improving the accuracy of identification. This capability aids in tracking outbreaks and identifying sources of contamination. Although powerful, these methods can be expensive and require sophisticated equipment and training to operate effectively.

Beyond bacteria and viruses, parasitic contamination is a significant concern in food safety. Biosensors are being developed for the detection of various parasites, including Cryptosporidium and Toxoplasma gondii. These often employ aptamer-based assays for recognition. Aptamers are short, single-stranded DNA or RNA molecules that can bind with high specificity to target molecules. Their use allows for the development of biosensors with high specificity and sensitivity for parasite detection, minimizing false positives. This reduces unnecessary product recalls and increases consumer confidence. De-

veloping reliable and affordable aptamer-based biosensors remains a challenge, particularly for detecting parasites at low concentrations in complex food samples.
The future of biosensor technology in foodborne pathogen detection lies in the continued development of portable, user-friendly, and cost-effective devices. The integration of nanotechnology, microfluidics, and advanced signal processing techniques will enhance sensitivity, specificity, and speed. This will allow for rapid on-site testing with minimal sample preparation, drastically reducing the time needed to identify and respond to contamination events. These advancements will transform food safety protocols, leading to safer food products and improved public health. The integration of biosensor data with advanced analytics, including machine learning and artificial intelligence (excluding these terms), could allow for predictive modeling of contamination risks. This would lead to preventative measures and proactive interventions, rather than solely reactive responses to outbreaks.

2.1 Bacterial Detection Methods

Rapid and accurate bacterial detection is paramount in ensuring food safety. Traditional methods, like culture-based techniques, are time-consuming and may not be sensitive enough for detecting low levels of contamination. Biosensors offer a powerful alternative, providing rapid, sensitive, and often portable solutions for detecting various bacterial pathogens in food products. Their use significantly reduces the time to results, allowing for quicker responses to potential contamination events and minimizing the risk of widespread outbreaks. Several biosensor types are particularly well-suited for bacterial detection. Electrochemical biosensors, for instance, measure changes in electrical current or potential associated with the presence of specific bacterial components. These biosensors often utilize antibodies or aptamers as recognition elements, binding specifically to target bacteria and triggering a measurable signal. The simplicity and relatively low cost of many electrochemical platforms make them attractive options for on-site testing, particularly in resource-limited settings.

Optical biosensors provide another powerful tool, leveraging changes in light properties, such as absorbance, fluorescence, or refractive index, to signal bacterial presence. Surface plasmon resonance (SPR) biosensors, a prime example, offer high sensitivity and label-free detection, providing real-time monitoring of bacterial binding events. Fluorescence-based methods often employ labeled antibodies or other probes that emit light upon binding to target bacteria, offering a visual indication of contamination levels. Their sensitivity and versatility contribute to their widespread use in research and industrial settings.

Piezoelectric biosensors exploit the changes in frequency of a quartz crystal upon binding of bacteria to its surface. This mass-sensitive detection method is particularly useful for detecting bacterial cells directly, without the need for labeling or signal amplification. Moreover, they offer a label-free detection, enabling faster analysis and reducing complexity. However, challenges in controlling surface chemistry for specific and sensitive target binding are ongoing areas of research and development.

Beyond these common techniques, various other methods and approaches are constantly being developed. Impedimetric biosensors measure changes in electrical impedance arising from bacterial growth on the electrode surface; this technique is well-suited for continuous monitoring of bacterial growth kinetics. Colorimetric biosensors use color changes to indicate the presence of bacteria, which may provide visual indication of contamination, suitable for simpler tests. The choice of a specific biosensor method for bacterial detection depends heavily on various factors. The target bacterium, the desired sensitivity and specificity, the sample matrix, and the available resources all play crucial roles in determining the most suitable approach. For example, detecting Salmonella in poultry samples might necessitate a different biosensor than detecting Listeria in ready-to-eat foods. These factors should be carefully considered during the selection and validation of a bacterial detection method for a specific

application.
The integration of biosensors into existing food safety protocols has the potential to revolutionize the industry. Rapid, on-site detection allows for immediate corrective actions, preventing the distribution of contaminated products and protecting consumers. However, widespread adoption necessitates addressing several challenges. These include ensuring the robustness and reliability of biosensors in complex food matrices, standardizing detection procedures, and integrating the technology into existing infrastructure. Further research and development are necessary to make biosensors more cost-effective and user-friendly, broadening their accessibility to a wider range of stakeholders. Addressing these challenges will make biosensor-based bacterial detection a staple in the realm of food safety management.

2.2 Viral Detection Strategies

Viral detection in food necessitates rapid, sensitive, and specific methods to prevent outbreaks. Current approaches leverage several biosensing platforms, each with unique strengths and limitations. Immunosensors, for example, utilize antibodies specific to viral surface antigens. These antibodies, immobilized on a transducer surface, bind to viral particles, triggering a measurable signal. Electrochemical immunosensors, measuring changes in current or potential, offer high sensitivity and are relatively inexpensive. Optical immunosensors, utilizing surface plasmon resonance or fluorescence, provide label-free detection and high throughput capabilities, ideal for large-scale screening. However, antibody production can be time-consuming and expensive, and cross-reactivity with other viruses may occur.
Nucleic acid-based biosensors offer a complementary strategy, targeting viral genetic material. These sensors often employ polymerase chain reaction (PCR) or its variations, dramatically amplifying viral DNA or RNA sequences for detection. PCR-based biosensors integrated with microfluidic devices enable rapid, automated analysis, while electrochemical and optical detection methods can further

enhance sensitivity and throughput. However, PCR's sensitivity to contamination requires stringent control and careful handling, adding complexity to the assay. The specificity of PCR is crucial, especially when targeting closely related viral strains. False positives or negatives can have significant consequences.

Aptamer-based biosensors utilize short, single-stranded DNA or RNA sequences that bind to specific viral targets. Aptamers offer several advantages over antibodies, including higher stability, easier modification, and lower production costs. Aptamer-based electrochemical and optical biosensors have shown promise in detecting various foodborne viruses. However, aptamer selection and optimization can be challenging and requires extensive screening to identify those with high affinity and specificity. Careful design and optimization are key to ensure reliable detection, minimizing false results.

Another promising technology is phage display. Phage display allows for the identification of peptides or proteins that specifically bind to viral targets. These peptides can then be used as recognition elements in biosensors, offering an alternative to antibodies or aptamers. This approach allows for more rapid screening of potential binding elements, offering cost-effective high-throughput capabilities. However, the inherent complexity of this method could prove to be problematic in developing commercial sensors for widespread application. This method demands careful validation and standardization to ascertain the accuracy and reliability of the results.

Furthermore, advancements in nanotechnology have led to the development of novel biosensors for viral detection. Nanomaterials, such as gold nanoparticles, quantum dots, and carbon nanotubes, enhance the sensitivity and selectivity of existing platforms by increasing the surface area for binding events, thus offering improved signal amplification. These nanoscale-based sensors are advantageous in providing a high signal-to-noise ratio and the capability for miniaturization for point-of-care applications. However, the potential toxicity of some nanomaterials raises safety concerns requiring careful consideration and regulation.

Integrating different detection strategies provides a comprehensive

approach to viral detection in food, optimizing the advantages of each method while mitigating their limitations. Combining antibody-based assays with nucleic acid-based techniques enhances sensitivity and specificity, ensuring accurate identification even in complex food matrices. Combining these techniques allows for a more robust and effective detection system, capable of distinguishing multiple viral targets. Integrating these into a portable device can improve food safety testing in diverse settings.
The development of portable, user-friendly, and cost-effective biosensors is critical for widespread adoption in food safety testing. Point-of-care devices, capable of providing rapid results on-site, can significantly reduce the time required for detection and enable immediate action to prevent outbreaks. Developing a simple format is crucial in widening the reach to improve quality checks. Such devices would empower food producers, processors, and regulatory agencies to proactively monitor food safety. However, integrating all these systems requires a comprehensive and multifaceted approach to technological innovation.
Standardization and validation of biosensor technologies are essential for ensuring reliable and consistent results. Developing standardized protocols and reference materials ensures that different laboratories obtain comparable results, promoting inter-laboratory agreement and compliance with regulatory standards. This increases the confidence in the results provided. However, the complexity of some of these procedures remains a challenging aspect, potentially hindering the wide acceptance of the technique. The cost-effectiveness needs to be improved to enable wider adoption.
Future research should focus on improving the sensitivity, specificity, and robustness of viral detection biosensors. This includes exploring new recognition elements, enhancing signal transduction methods, and developing advanced data analysis techniques. The incorporation of machine learning algorithms could further improve the accuracy and speed of viral detection. Integrating artificial intelligence could aid in generating high-quality, accurate data for further analysis, especially in complex food matrices. Developing more rapid,

simple, and reliable methods would facilitate effective food safety measures globally.

2.3 Parasitic Contamination Analysis

Parasitic contamination in food poses a significant threat to public health, demanding rapid and sensitive detection methods. Traditional methods, often laborious and time-consuming, are insufficient for the speed and scale required by modern food production and distribution. Biosensors offer a powerful alternative, providing rapid, on-site detection with high sensitivity and specificity. The inherent ability of biosensors to directly target specific parasitic antigens or nucleic acids is particularly advantageous, leading to a more efficient and accurate diagnostic process compared to more generalized techniques.

Electrochemical biosensors, for instance, are particularly well-suited for detecting parasitic contamination. These devices leverage the specific binding of antibodies or aptamers to parasite antigens, generating measurable electrical signals proportional to the parasite concentration. The simplicity and relatively low cost of electrochemical biosensors make them attractive for widespread adoption, especially in resource-limited settings where rapid detection is crucial for outbreak management. Miniaturization of these devices is an ongoing area of research, further enhancing portability and ease of use. Imagine a handheld device capable of quickly assessing food samples in the field, providing immediate results to prevent contaminated products from entering the food chain.

Optical biosensors represent another powerful tool in the fight against parasitic contamination. These sensors rely on changes in optical properties, such as absorbance, fluorescence, or refractive index, to signal the presence of parasites. Surface plasmon resonance (SPR) biosensors, for example, offer high sensitivity and label-free detection, directly measuring the interaction between parasite antigens and their corresponding antibodies. Furthermore, the use of fluorescently labeled probes coupled with microfluidic systems allows for high-throughput screening and precise quantification of parasitic load in food samples. This technology offers the capability to process numerous samples simultaneously, accelerating the analytical process significantly.

Impedimetric biosensors offer a unique approach to parasite detection, focusing on changes in electrical impedance caused by the binding of parasitic antigens to immobilized receptors on the sensor surface. The simplicity of the technology, coupled with its high sensitivity, makes impedimetric biosensors a promising approach, particularly for detecting low levels of parasitic contamination often found in processed foods. This technology's ability to provide quantitative data on the extent of contamination is particularly valuable for risk assessment and regulatory compliance. Further development focuses on improving the robustness and reliability of these sensors to withstand the harsh conditions sometimes present in food processing environments.

Developing robust and reliable biosensors for parasitic detection demands careful consideration of various factors. The specificity of the biorecognition element is paramount, ensuring that the sensor only detects the target parasite and minimizes false positives from other food components. Sensitivity is equally crucial, particularly for detecting low levels of contamination that may not trigger visible symptoms but can still cause serious health consequences. Furthermore, the sensor needs to be stable and reliable over a range of operating conditions such as temperature and humidity, ensuring consistent and accurate results. Robustness in the face of food matrix interference is also crucial for effective deployment in real-world applications.

Data analysis and interpretation play a pivotal role in the successful application of biosensors for parasitic contamination analysis. Data processing algorithms are essential for removing noise, identifying true signals, and accurately quantifying parasite levels. Calibration of the sensors using known standards is vital for ensuring accuracy and reliability. Robust quality control protocols are equally important, ensuring the reproducibility and validity of the results. Furthermore, appropriate statistical methods are needed to analyze the data and interpret results with confidence, providing a clear understanding of the extent of parasitic contamination in food samples. The development of user-friendly software tools to facilitate these processes would greatly benefit the wider adoption of biosensor technologies in the food industry.

Beyond the technical aspects, regulatory considerations and standardization are crucial for the successful implementation of biosensor technologies in food safety. Establishing clear guidelines and protocols for the validation and approval of biosensors is essential to ensure the reliability and acceptance of the results. The development of internationally recognized standards for biosensor performance will also streamline their adoption by various food companies and regulatory bodies. This would facilitate the sharing of data, improve comparability of results across different labs, and strengthen the overall confidence in the accuracy and reliability of biosensor-based detection methods. Collaboration between researchers, food companies, and regulatory agencies is paramount to overcoming these challenges and paving the way for widespread adoption of these powerful technologies.

3 Biosensors for Food Toxin Detection

Mycotoxins, naturally occurring toxins produced by fungi, pose a significant threat to food safety. Aflatoxins, a particularly dangerous group of mycotoxins, contaminate various food commodities like grains, nuts, and spices, causing severe health problems, including liver cancer. Biosensors offer a rapid, sensitive, and cost-effective solution for detecting aflatoxins in these food products. Electrochemical biosensors, for instance, utilize antibodies or aptamers as recognition elements to bind specifically to aflatoxins. The binding event is then transduced into an electrical signal, providing a quantifiable measure of aflatoxin concentration. This method allows for on-site testing, reducing the time required for laboratory analysis and facilitating rapid responses to contamination events. The development of portable and user-friendly electrochemical biosensors is crucial for effective implementation in resource-limited settings, empowering local producers and regulators.

Beyond aflatoxins, other mycotoxins like ochratoxins and fumonisins require equally robust detection methods. Immunoassays coupled with biosensor platforms provide highly sensitive and specific detection capabilities. These assays leverage the high affinity between antibodies and target mycotoxins, converting the immunological reaction into a measurable signal using various transduction mechanisms, such as fluorescence or colorimetric changes. The sensitivity of these biosensors is crucial for detecting low levels of contamination, below the regulatory limits established for food safety. Furthermore, the development of multiplexed biosensor platforms enables the simultaneous detection of various mycotoxins within a single sample, increasing efficiency and reducing analysis costs. This approach is especially important for complex food matrices where multiple mycotoxins may co-occur.

Pesticide residues in food products are a significant concern for public health. Many conventional pesticides persist in the environment and can accumulate in food crops, causing adverse health effects. Biosensors offer a powerful tool for monitoring pesticide residues. Enzymatic biosensors, for example, exploit the high specificity of enzymes toward certain pesticides. The enzyme catalyzes a reaction only in the presence of the target pesticide, producing a detectable signal proportional to the concentration. However, these methods must be carefully calibrated and validated. Furthermore, the development of multi-analyte biosensors for detecting various pesticide residues is critical for comprehensive food safety assessments. This would enable regulatory agencies to monitor various pesticides simultaneously.

Heavy metal contamination in food poses another serious threat, stemming from various sources including industrial pollution and agricultural practices. Heavy metals like lead, mercury, and cadmium accumulate in the food chain, causing significant health issues. Biosensors provide a solution for rapid and sensitive detection. Optical biosensors, often employing fluorescent probes, can be utilized to detect heavy metal ions in food samples. These probes change their fluorescence intensity upon binding to heavy metals, which then permits for easy quantification. The selectivity of these sensors is crucial to avoid false positives from other metal ions present in the food matrices, hence precise calibration and rigorous testing are essential steps in the development of these biosensors. Research focuses on enhancing the selectivity and sensitivity of these sensors, enabling their widespread use in food safety monitoring programs.

The application of biosensors in food toxin detection is rapidly evolving. The need for rapid, sensitive, and cost-effective methods for detecting foodborne toxins is paramount, driving innovation in biosensor design and development. Technological advancements, such as nanotechnology and microfluidics, further enhance the performance and practicality of biosensors. Nanomaterials offer unique advantages, including increased surface area for improved sensitivity and the potential for miniaturization, making portable and point-of-care diagnostics a reality. However, the translation of these advancements into commercially viable and widely adopted technologies requires addressing certain challenges. Ensuring the reproducibility and long-term stability of biosensors is crucial for their reliable performance in real-world applications. Standardization of biosensor protocols and validation procedures is also necessary for ensuring data comparability and acceptance by regulatory authorities. Overcoming these hurdles is essential for realizing the full potential of biosensors in safeguarding food safety and protecting public health.

3.1 Mycotoxin Detection

Mycotoxins, toxic secondary metabolites produced by various fungi, pose significant threats to food safety and human health. Their presence in food products can lead to a range of adverse effects, from mild gastrointestinal distress to severe liver damage and even cancer. Effective detection methods are therefore crucial for protecting consumers and ensuring the safety of the food supply chain. Traditional methods, like High-Performance Liquid Chromatography (HPLC) and Gas Chromatography-Mass Spectrometry (GC-MS), while accurate, are often time-consuming, expensive, and require specialized equipment and trained personnel. This necessitates the development of rapid, sensitive, and cost-effective alternatives. Biosensors offer a promising solution, providing a platform for detecting mycotoxins with enhanced speed and efficiency compared to traditional methods.

The development of biosensors for mycotoxin detection hinges on the selection of a suitable biorecognition element. Antibodies, aptamers, and enzymes all display strong potential. Antibodies, for example, exhibit high specificity and affinity for target mycotoxins, making them ideal candidates for immunoassays. Aptamers, single-stranded DNA or RNA molecules, offer advantages such as stability, easy synthesis, and cost-effectiveness, representing a valuable alternative. Enzymes, owing to their catalytic abilities, can be integrated into biosensors to provide signal amplification and enhance sensitivity. The choice of biorecognition element often depends on the specific mycotoxin being targeted, as well as the desired sensitivity and detection limit. Careful consideration of these factors is paramount for the successful development of a robust and reliable biosensor.

Transducer selection is another critical aspect of biosensor design. Electrochemical transducers, which measure changes in electrical current or potential, are commonly employed due to their simplicity, sensitivity, and affordability. Optical transducers, based on changes in light absorption or fluorescence, offer excellent sensitivity and the ability to perform measurements in real-time, providing continuous monitoring capabilities. Piezoelectric transducers, measuring changes in mass or frequency, are particularly suitable for detecting changes caused by binding interactions between mycotoxins and biorecognition elements, allowing for highly sensitive and specific detection. The most appropriate choice often involves considering factors such as cost, ease of use, and the specific detection requirements of the mycotoxin under investigation.

Beyond the core components, the overall design and optimization of the biosensor system are essential for achieving optimal performance. This involves careful selection of immobilization methods to ensure the stability and activity of the biorecognition element. Microfabrication techniques, which allow for miniaturization and integration of multiple components onto a single chip, are increasingly being used to develop portable and user-friendly devices. Moreover, careful optimization of assay parameters, such as incubation time and temperature, is crucial for maximizing sensitivity and reducing false-positive or false-negative results. A well-designed biosensor should exhibit high specificity, sensitivity, and reproducibility, ensuring accurate and reliable detection of mycotoxins in various food matrices.

The development and application of biosensors for mycotoxin detection is a rapidly evolving field. Recent advances in nanotechnology and materials science have led to the creation of novel biosensors with enhanced sensitivity, portability, and cost-effectiveness. The integration of nanomaterials, such as nanoparticles and carbon nanotubes, into biosensor designs can improve the signal transduction efficiency and expand the range of detectable mycotoxins. Furthermore, the development of portable and point-of-care biosensors enables on-site detection, reducing the time and cost associated with laboratory testing, allowing for timely intervention and prevention of mycotoxin contamination. Continuous research is needed to further refine and optimize these technologies for wider adoption in the food industry, ultimately contributing to a safer food supply for consumers worldwide. The ultimate goal remains a fully integrated system that is rapid, accurate, and affordable enough for widespread use in quality control settings.

3.2 Aflatoxin Analysis

Aflatoxins, potent mycotoxins produced by Aspergillus species, pose a significant threat to food safety globally. Their presence in staple crops like maize, peanuts, and tree nuts can lead to serious health consequences, ranging from liver damage to cancer. Accurate and rapid detection is, therefore, paramount. Traditional methods, while effective, often prove time-consuming and require specialized laboratory equipment. This necessitates the development of more accessible and efficient detection techniques.

Biosensors offer a promising solution. Their inherent specificity and sensitivity enable the rapid and reliable detection of aflatoxins even at very low concentrations. Various biosensor platforms have been explored, each presenting unique advantages and limitations. Electrochemical biosensors, for instance, exploit the changes in electrical signals upon aflatoxin binding to a specific receptor. This method is relatively simple, cost-effective, and amenable to miniaturization for on-site testing. However, the sensitivity might be limited depending

on the chosen receptor and the detection system's efficiency. Immunosensors, utilizing antibodies as recognition elements, offer high specificity, but production and maintenance costs can be substantial. The choice of biosensor platform depends largely on the intended application, available resources, and the required level of sensitivity.
Aptamers, single-stranded DNA or RNA molecules that bind specifically to target molecules, offer a compelling alternative to antibodies. They are relatively easy to synthesize, highly stable, and can be modified to enhance their affinity and selectivity. Aptamer-based biosensors have shown impressive results in aflatoxin detection, offering a potentially more cost-effective and readily adaptable solution. Research into novel aptamer sequences continues to refine their capabilities, pushing towards even greater sensitivity and specificity. These improvements could lead to more reliable and readily available point-of-care diagnostic tests, directly impacting food safety measures in resource-limited settings.
Optical biosensors provide another avenue for aflatoxin detection. Techniques like surface plasmon resonance (SPR) and fluorescence spectroscopy offer highly sensitive detection methods. SPR measures changes in the refractive index at a metal-dielectric interface upon aflatoxin binding, providing a label-free detection method. Fluorescence spectroscopy uses fluorescently labeled antibodies or aptamers to quantify aflatoxin levels. While offering excellent sensitivity, these techniques can be relatively complex and expensive, limiting their widespread adoption. This higher level of technological sophistication often leads to the requirement of experienced operators, which could limit accessibility in some regions.
The integration of biosensors with microfluidic devices presents exciting opportunities for miniaturization and automation of aflatoxin analysis. Microfluidic devices can precisely control fluid flow, allowing for efficient sample processing and reducing the volume of reagents needed. Combining this with electrochemical or optical biosensors could lead to portable and user-friendly devices for on-site testing. This integration also presents the potential for higher throughput analysis, significantly increasing the number of samples

that can be processed. This high throughput testing is critical when conducting large-scale surveillance or quality control for food manufacturers and regulatory bodies.
Furthermore, advancements in nanotechnology are enhancing biosensor performance. Nanomaterials, such as gold nanoparticles and carbon nanotubes, can be incorporated into biosensor designs to improve sensitivity and signal amplification. These nanomaterials can enhance the efficiency of the detection systems, potentially leading to more rapid and sensitive aflatoxin analysis. Their increased surface area facilitates the binding of more aflatoxin molecules, resulting in a stronger signal and improved sensitivity. However, careful consideration must be given to the potential toxicity and environmental impact of these nanomaterials.
Despite the significant advances in aflatoxin biosensor technology, several challenges remain. Matrix effects from food samples can interfere with detection, requiring careful sample preparation. The need for standardization and validation of biosensor assays is crucial for ensuring accurate and reliable results. The development of cost-effective and user-friendly devices is essential for widespread adoption, especially in developing countries where aflatoxin contamination poses a significant risk. Continued research efforts focused on addressing these challenges are vital for the successful implementation of biosensors in improving global food safety and security. Overcoming these hurdles promises to revolutionize aflatoxin detection, creating a safer and more reliable food supply chain for all.

3.3 Other Toxin Detection

Beyond aflatoxins and mycotoxins, a wide array of other toxins threaten food safety. These contaminants, originating from various sources, demand equally diverse detection methods. Think about the impact of ochratoxin A, a mycotoxin produced by Aspergillus and Penicillium species, frequently found in cereals, coffee, and wine. Its nephrotoxic effects necessitate stringent monitoring, often achieved through techniques like high-performance liquid chromatography (HPLC) coupled with fluorescence or mass spectrometry detection, offering unparalleled sensitivity and precision. These sophisticated methods, while effective, can be expensive and require specialized equipment, posing a barrier for smaller food production facilities.

However, innovative biosensor technologies are emerging to address this challenge. Aptamer-based biosensors, for instance, provide a cost-effective and rapid alternative. Aptamers, single-stranded DNA or RNA molecules, are selected for their high affinity and specificity towards target molecules, including ochratoxin A. These aptamers can be immobilized on a sensor surface, triggering a measurable signal upon binding the toxin. This signal's intensity is directly proportional to the toxin's concentration, providing a quantitative measure. The ease of production, coupled with the possibility of on-site testing, makes this approach particularly attractive for routine quality checks in diverse settings.

Another critical concern is the presence of biogenic amines in food. These nitrogenous compounds, formed through microbial decarboxylation of amino acids, can cause adverse health effects ranging from headaches and nausea to more severe reactions. Histamine, a common biogenic amine found in fermented foods like fish and cheese, is a notorious culprit. Traditional methods for histamine detection involve lengthy chemical analyses, but electrochemical biosensors offer a streamlined approach. Enzymes specific to histamine can be incorporated into the sensor design, converting the amine into a detectable signal. The rapidity and portability of such electrochemical systems allow rapid on-site screening of products, preventing contaminated batches from reaching consumers.

Moving beyond microbial toxins, we must also consider the presence of marine biotoxins. These toxins, primarily associated with shellfish, pose a significant risk to human health. Saxitoxin, a potent neurotoxin produced by certain dinoflagellates, can lead to paralytic shellfish poisoning. Currently, the most common detection methods rely on sophisticated techniques like liquid chromatography-mass spectrometry, while immunoassay based approaches are also employed.

However, emerging biosensors based on specialized antibodies or aptamers that target specific marine toxins are gaining traction. These developments promise to deliver faster, less expensive detection tools for use in field testing scenarios, which is crucial given the rapid and unpredictable nature of marine toxin contamination. The sensitivity and specificity of these emerging technologies will be paramount in protecting public health.

Moreover, the detection of cyanotoxins, toxins released by cyanobacteria, is a critical aspect of food and water safety. Microcystins, a family of hepatotoxins, represent a significant health concern. Traditional analytical methods, including high-performance liquid chromatography (HPLC) and enzyme-linked immunosorbent assays (ELISA), are employed, but they are often time-consuming and require specialized facilities. The development of portable biosensors that can provide rapid and accurate detection of microcystins on-site has the potential to greatly enhance food and water safety protocols, ensuring rapid responses to outbreaks and minimizing potential human health risks. The integration of nanomaterials into these biosensor platforms further enhances their sensitivity and functionality, making them potentially ideal tools for real-world applications.

Finally, the impact of environmental pollutants finding their way into the food chain cannot be overlooked. Pesticide residues, while often present in low concentrations, represent a significant concern. Existing methods for detection often involve complex extraction and purification procedures, typically requiring sophisticated laboratory equipment. Biosensors, especially those based on antibody-antigen interactions or aptamers, offer the potential for faster, simpler, and more cost-effective alternatives. Researchers are working on developing sensors that use specific binding molecules for diverse pesticide types, leading to a more comprehensive system for evaluating the safety of food products. The development of portable devices will have a direct and substantial effect on the speed at which these analyses can be conducted. This would significantly improve the rapid response potential in situations of suspected pesticide contamination.

3.4 Pesticide Residue Analysis

Pesticide residue analysis in food presents a significant challenge to ensuring food safety. Effective detection methods are crucial, not only for regulatory compliance but also for protecting public health. The potential for long-term health consequences associated with pesticide exposure underscores the importance of sensitive and reliable analytical techniques. Current methods, while effective, often involve complex procedures and expensive instrumentation, potentially limiting their accessibility, particularly in resource-constrained settings. Developing rapid, cost-effective, and user-friendly methods for pesticide residue detection is therefore essential. Biosensors offer a promising avenue to address this need. Their inherent advantages—such as high sensitivity, specificity, and portability—make them attractive alternatives to traditional chromatographic and spectroscopic techniques. The ability to miniaturize biosensors further enhances their potential for on-site analysis, facilitating quick decision-making in food safety assessments. Furthermore, biosensors can be designed for multiplexed detection, allowing for the simultaneous analysis of various pesticide residues in a single sample.

Several biosensor types show significant promise in pesticide residue analysis. Electrochemical biosensors, for instance, leverage the changes in electrical signals upon interaction of pesticides with specific biological receptors. These receptors, often enzymes or antibodies, are immobilized on an electrode surface. Enzyme-linked immunosorbent assays (ELISAs), adapted for biosensor platforms, offer high sensitivity and specificity. Optical biosensors utilize changes in light properties—such as absorbance, fluorescence, or reflectance—to detect pesticide-receptor interactions. Surface plasmon resonance (SPR) and fiber optic sensors are examples of this category. Piezoelectric biosensors, based on the change in mass upon binding of pesticide molecules, represent another viable option, offering advantages in sensitivity and simplicity.

The choice of biosensor type depends on several factors, including the target pesticide, the required sensitivity, the availability of appropriate bioreceptors, and the cost-effectiveness of the overall system. For instance, the selection of antibodies or aptamers as bioreceptors dictates the specificity of the assay. Careful optimization of parameters like immobilization strategies, signal transduction mechanisms, and detection limits is critical for achieving reliable and accurate results. The matrix effect, which arises from the complex nature of food samples, needs to be considered and mitigated using appropriate sample preparation techniques.

While significant advancements have been made, challenges still persist. The development of robust and stable bioreceptors with high affinity and specificity for various pesticide residues remains an ongoing effort. Ensuring the long-term stability of the biosensor devices, particularly in field conditions, presents another challenge. Furthermore, validation and standardization of biosensor-based analytical methods are essential for their widespread acceptance and regulatory approval. Extensive inter-laboratory comparisons and collaborative studies are needed to establish reliable performance parameters. Finally, addressing the cost-effectiveness and ease of use of biosensor systems are crucial factors for their successful implementation in routine food safety monitoring.

Moreover, integrating biosensors into existing food safety workflows requires careful consideration. The integration process needs to be seamless, minimizing disruptions to established protocols. Training personnel to operate and maintain biosensors is also essential for successful implementation. Ultimately, the widespread adoption of biosensors for pesticide residue analysis relies not only on technological advancements but also on effective strategies for training, implementation, and regulatory support. The development of user-friendly interfaces, coupled with comprehensive training programs, will be crucial in facilitating wider acceptance and utilization of these technologies in diverse settings.

The future of pesticide residue analysis hinges on the continued development of innovative biosensor technologies, enhanced by advancements in nanotechnology, microfluidics, and materials science. The integration of artificial intelligence and machine learning algorithms can further improve data analysis and interpretation. In addition, ongoing research efforts focused on creating more robust and stable biosensors that can withstand harsh environmental conditions and prolonged storage will contribute greatly to the overall adoption. Furthermore, addressing the cost factor and ensuring affordability will be vital to creating wide access to this crucial technology. Ultimately, the goal is to create a simple, effective, and reliable system that empowers individuals to make informed choices regarding their food choices.

3.5 Heavy Metal Detection

Heavy metal contamination poses a significant threat to food safety and human health. These metals, unlike organic contaminants, are not metabolized by the body and accumulate over time, leading to various health problems, including neurological disorders and cancers. Therefore, sensitive and reliable detection methods are crucial for ensuring food safety and protecting consumers. Existing techniques, while effective, often require extensive sample preparation, sophisticated instrumentation, and trained personnel, limiting their widespread application, especially in resource-constrained settings.

Biosensors offer a powerful alternative, providing a rapid, sensitive, and cost-effective approach to heavy metal detection in food. These miniaturized analytical devices combine a biological recognition element, such as an antibody or enzyme, with a transducer that converts the biological interaction into a measurable signal. The choice of recognition element and transducer dictates the specificity and sensitivity of the biosensor. For instance, an antibody-based biosensor can specifically target a particular heavy metal ion, while an electrochemical transducer can measure the resulting change in electrical current. This direct, highly selective approach simplifies the analytical process.

Several types of biosensors have shown great promise in detecting heavy metals. Electrochemical biosensors, for example, are often favored for their simplicity, cost-effectiveness, and portability. They rely on changes in current or potential caused by the interaction of the heavy metal with the biosensor's biological recognition element. Optical biosensors, based on changes in light absorption or fluorescence, offer high sensitivity and can be easily integrated with other detection systems. These technologies allow for real-time monitoring and even on-site analysis, greatly enhancing the speed and accessibility of heavy metal detection in food.

The design and optimization of these biosensors is a crucial step toward their successful application. Factors such as the choice of biological recognition element, the immobilization strategy, the transducer type, and the signal processing method all significantly affect the biosensor's performance characteristics. Careful selection and optimization of these factors are critical for ensuring high sensitivity, specificity, and reproducibility. Moreover, the development of robust and stable biosensors capable of withstanding harsh conditions is essential for practical applications in real-world food matrices.

Besides the technical challenges, the broader context of implementing these technologies in food safety settings also requires attention. The cost-effectiveness of the biosensor system, its ease of use, and the availability of appropriate training for personnel are paramount factors determining its successful integration into food quality control procedures. Furthermore, the development of standardized protocols and regulations is needed to ensure the accuracy and reliability of biosensor-based heavy metal detection. Regulatory approval is essential for the widespread adoption of any novel food safety technology, ensuring consumer confidence and protecting public health.

The successful implementation of biosensor-based heavy metal detection requires a multidisciplinary approach, bringing together expertise in biochemistry, analytical chemistry, engineering, and regulatory affairs. Collaboration among researchers, industry stakeholders, and regulatory bodies is critical to overcoming the challenges and promoting the wider adoption of these technologies. By embracing these strategies, the biosensor community can play a significant role in improving food safety and securing global food supplies. The development of user-friendly, cost-effective biosensors will be transformative, facilitating widespread testing and potentially preventing countless cases of heavy metal poisoning. This, ultimately, is the goal: safeguarding public health through technological innovation. The potential of biosensors in heavy metal detection is enormous, offering a significant step towards ensuring the safety and quality of our food supply. Through continued research and development, these devices promise a future where heavy metal contamination is swiftly detected, and appropriate action can be taken to protect consumers. This is a crucial step towards a safer food supply chain, benefiting both the industry and the public.

4 Monitoring Food Quality Parameters

Assessing food freshness is crucial for minimizing waste and ensuring consumer satisfaction. Traditional methods, often reliant on sensory evaluation or time-consuming laboratory analyses, lack the speed and precision demanded by modern food supply chains. Biosensors offer a powerful alternative, providing rapid, non-destructive, and objective measurements of key freshness indicators. For example, biosensors can detect volatile organic compounds (VOCs) released during spoilage, providing an early warning signal before perceptible changes occur. The concentration and type of VOCs, specific to particular spoilage microorganisms, can be used to create a detailed profile of the food's freshness level. This allows for more precise predictions of shelf life, leading to optimized inventory management and reduced food loss. Imagine a system that can instantly assess the quality of a shipment of produce upon arrival, identifying potentially spoiled items before they even reach the shelves. That's the power of biosensing technology in action, directly translating to significant financial savings and enhanced consumer confidence.

Analyzing the nutritional content of food is equally important. Consumers are increasingly aware of the nutritional value of their diet, driving demand for accurate and readily accessible nutritional information. Traditional methods for analyzing nutrient content are often expensive, time-consuming, and require specialized laboratory equipment. Biosensors provide a more efficient approach, enabling rapid and simultaneous determination of multiple nutrients. For instance, electrochemical biosensors can measure the concentration of vitamins and minerals, while optical biosensors can quantify the levels of specific carbohydrates or proteins. These portable biosensors can be used directly at the point of production, processing, or even at the retail level, providing immediate feedback on the nutritional composition of food products. This direct, on-the-spot data empowers producers to make informed decisions about quality control, optimize their processes for consistent nutrient levels, and develop more nutritionally enriched products. This fosters transparency and boosts consumer trust in product claims.

Beyond simple detection, the integration of biosensors into more comprehensive systems provides a holistic approach to food quality monitoring. By combining data from various sensors, it's possible to build predictive models for food shelf life and quality. These models take into account multiple parameters, including temperature, humidity, and the presence of spoilage indicators, to provide a more accurate and reliable estimate of remaining shelf life. This enhanced predictive capability allows for proactive measures to maintain quality, such as adjusting storage conditions or implementing targeted interventions. Furthermore, using a network of interconnected sensors, one can monitor food quality across the entire supply chain, from farm to table. This enables early identification of quality issues, allowing for timely intervention and preventing large-scale losses.

The use of biosensors in food quality monitoring isn't limited to just the detection and quantification of specific compounds. Advanced biosensors are being developed which can identify complex mixtures of molecules, providing a more comprehensive analysis of food quality. For example, sophisticated sensors based on advanced materials, like nano-materials and polymers, can detect subtle changes in food composition indicative of early spoilage or oxidation. The application of this technology can improve decision-making regarding optimal harvest times, processing conditions, and storage procedures, thereby minimizing waste and ensuring the delivery of high-quality food products to the consumer. Real-time monitoring, made possi-

ble through interconnected sensor networks and sophisticated data analysis techniques, offers the potential for significant changes to the way the food industry operates, making it more efficient, sustainable and consumer-focused. The development and implementation of these tools are not merely technological improvements but represent a paradigm shift in how we approach food safety and quality assurance.
The development of user-friendly biosensors is paramount for widespread adoption. Complex instrumentation and specialized expertise can be barriers to implementation, especially for small-scale producers or in developing countries. The creation of simple-to-use, portable biosensors, requiring minimal training, is crucial to making this technology accessible to a wider range of stakeholders. This accessibility translates into broader implementation of food quality control measures, resulting in greater food security and enhanced public health. Ultimately, the goal is to create biosensor systems which require minimal technical expertise while providing maximum functionality, effectively empowering even small-scale food producers to implement superior quality control practices. Such a system would lead to better quality food across the board, positively impacting both producers and consumers.

4.1 Assessing Freshness

Maintaining optimal food quality relies heavily on accurate and timely freshness assessment. Consumers prioritize fresh produce, and spoilage detection directly impacts food safety, preventing illnesses caused by bacterial growth or toxin production. Traditional methods, while effective, often prove slow and cumbersome, necessitating rapid, accurate alternatives. Biosensors offer precisely that: a revolutionary approach to food freshness assessment.
Rapid detection of volatile organic compounds (VOCs) released during spoilage is key. These compounds, indicative of microbial activity and enzymatic degradation, serve as reliable freshness markers. Electronic noses, for instance, employing arrays of gas sensors, offer

real-time detection and profiling of these VOCs, providing quantitative data correlating directly to the food's freshness. The data then undergoes sophisticated pattern recognition algorithms, generating a precise freshness score. This rapid analysis allows for immediate quality checks at various stages, from the farm to the supermarket shelf. Furthermore, these technologies minimize waste by pinpointing products nearing spoilage, enabling efficient stock management and targeted pricing strategies.

Beyond VOC analysis, other biosensors measure changes in relevant biochemical parameters. Enzyme activity, for example, diminishes with time, particularly those involved in metabolic processes. Biosensors targeting specific enzymes can therefore provide a reliable measure of freshness. Similarly, the pH value of a food product changes significantly during spoilage. Electrochemical biosensors specifically designed to measure pH changes can offer another crucial freshness indicator. The precision and speed offered by these technologies surpass traditional methods, especially visual inspection which relies heavily on subjective assessment and can miss early stages of deterioration. The combination of different sensor types offers a multifaceted approach, corroborating results for greater accuracy.

The application extends beyond simple freshness determination. Biosensors can provide granular insight into the underlying mechanisms of deterioration. By identifying the specific microorganisms or enzymes causing spoilage, we gain a deeper understanding of the degradation processes. This insight allows the development of targeted preservation strategies tailored to specific food types and their vulnerability to particular spoilage pathways. This strategic approach enables a significant improvement in food shelf life. Furthermore, the real-time monitoring facilitated by biosensors enhances traceability and strengthens quality control measures, enhancing overall safety and integrity across the food supply chain.

The integration of these advanced technologies into existing food handling infrastructure necessitates careful consideration. Ease of use and cost-effectiveness are paramount for widespread adoption. The development of portable, user-friendly devices, combined with streamlined data analysis software, is crucial to ensure practical application across all levels of the food industry. Moreover, standardized protocols for data acquisition and interpretation are necessary to ensure consistency and comparability of results. Standardization creates a platform for reliable communication and collaborative data analysis across different locations and organizations. This not only facilitates more effective research but enables a more integrated approach to food safety management across the global food supply chain. The long-term vision involves the development of interconnected networks capable of monitoring freshness and quality in real time across the entire food journey, from farm to consumer. Addressing these practical concerns enables the transition from laboratory experimentation to the widespread adoption of these powerful tools, significantly enhancing food safety and quality control.

Finally, the impact extends beyond economic gains. Accurate, rapid freshness assessment significantly reduces food waste, a pressing global concern. Decreased food waste translates to lowered environmental impact, reducing the strain on resources and lessening the carbon footprint associated with food production and distribution. By minimizing spoilage and extending shelf life, we contribute to a more sustainable and efficient food system. The resulting improvements across the food value chain, from economic viability to environmental sustainability, underscore the profound importance of incorporating biosensor technologies into modern food safety and quality control.

4.2 Analyzing Nutritional Content

Analyzing nutritional content in food is crucial for maintaining public health and ensuring product quality. Biosensors offer a rapid and precise approach to this challenge, moving beyond traditional methods that are often time-consuming and labor-intensive. We can leverage biosensors to detect specific vitamins, minerals, and other essential nutrients, providing consumers with detailed compositional information. This allows for better informed dietary choices and helps in addressing nutritional deficiencies effectively. The development of rapid, point-of-care diagnostic tools using biosensors significantly enhances the accessibility and feasibility of nutritional analysis, especially in resource-limited settings where access to sophisticated laboratory equipment may be lacking. This contributes to global food security and public health initiatives by empowering consumers and food safety agencies with real-time data.

Electrochemical biosensors, for instance, show great potential for quantifying vitamins. These devices use enzymes to catalyze specific reactions involving the targeted vitamins, generating measurable electrical signals proportional to the vitamin concentration. The simplicity and relatively low cost of these devices make them particularly attractive for widespread application. Further advancements in electrochemical biosensor technology are expected to enhance sensitivity and selectivity, allowing for the simultaneous detection of multiple vitamins within a single sample, thereby streamlining analysis. Research is also underway to improve portability and user-friendliness, enhancing suitability for diverse application settings, including home use. For example, miniaturized electrochemical biosensors are being integrated into portable devices for point-of-care testing, ensuring the rapid determination of nutritional values in various food products with minimal training required for operation.

Optical biosensors offer another powerful approach. Techniques like fluorescence spectroscopy and surface plasmon resonance can detect the presence and quantify specific nutrients through their interaction with labeled antibodies or other biorecognition elements. The high sensitivity and specificity of these techniques allow for the accurate determination of even trace amounts of essential nutrients. Furthermore, advancements in optical sensing technology allow for the development of fiber optic sensors which enable non-destructive analysis of food samples. This characteristic is advantageous since it avoids any alteration or damage to the food item during testing, maintaining the integrity of the sample for further analysis or consumption. Multiplexing capabilities are also being incorporated into optical biosensor platforms, enabling the simultaneous detection of multiple nutrients in a single test, reducing the overall analysis time significantly. These aspects underscore the value and efficiency in real-world applications in the food industry.

Beyond vitamins, biosensors are instrumental in analyzing the mineral content of food products. Minerals such as iron, zinc, and calcium are vital for numerous metabolic processes and their deficiency leads to various health issues. Biosensors based on ion-selective electrodes (ISEs) are highly effective in detecting various metal ions present in food samples. These devices utilize membranes that selectively bind to the target ions, producing an electrical potential that is directly proportional to the concentration of the ion. The rapid response time and relative ease of manufacturing are significant advantages. Ongoing research focuses on enhancing the selectivity of these electrodes to reduce interference from other ions that might be present in complex food matrices, a key step towards obtaining highly reliable quantitative data. This enhances confidence in data obtained from food analyses, increasing the accuracy and reliability of nutritional assessments.

However, the effective application of biosensors in nutritional analysis isn't without challenges. Matrix effects, which refer to the influence of non-target components in a sample, can significantly impact the accuracy of measurements. This necessitates the development of sophisticated sample preparation techniques and advanced data analysis algorithms to effectively mitigate interference from interfering substances and ensure accurate quantification of the target nutrients. Standardization of biosensor protocols across different laboratories is crucial for the reproducibility and reliability of results. This ensures the consistency and comparability of data obtained from diverse settings and allows for the establishment of benchmark values for the nutritional content of foods. Establishing stringent quality control measures and collaborative efforts between researchers, regulatory bodies and the food industry are necessary to improve reliability and acceptance.

Furthermore, the cost-effectiveness and ease of use are significant factors influencing the widespread adoption of biosensors for nutritional analysis. Researchers are actively developing low-cost, disposable biosensors that minimize the overall cost of testing. This aspect enhances the accessibility of such technology across different regions and populations. Simultaneously, efforts are being directed toward designing user-friendly biosensors which require minimal technical expertise for operation. The development of simple and intuitive interfaces is particularly crucial for improving accessibility for consumers, ensuring greater utilization of the technology and easier interpretation of nutritional data. These factors are critical to translating research findings into practical applications with widespread societal benefit. In conclusion, biosensors represent a powerful toolkit to revolutionize nutritional analysis.

5 Biosensor Applications in Food Processing

Real-time monitoring during food processing offers significant advantages. Imagine a system that continuously analyzes the levels of key metabolites or microbial contaminants, providing immediate feedback to optimize conditions. This proactive approach allows for adjustments to temperature, pressure, or other parameters, preventing spoilage and ensuring consistent product quality. Such systems can be implemented at various stages, from raw material inspection to final product packaging, improving overall efficiency and minimizing waste. This is not merely theoretical; several commercially available biosensors already facilitate this kind of real-time analysis, enhancing safety and efficiency. The data gathered can be integrated into sophisticated process control systems, enabling automated responses and reducing reliance on manual intervention. For example, in fermentation processes, biosensors can track the production of desired compounds or the growth of undesirable microorganisms, guiding adjustments to fermentation parameters for optimal yield and quality. This results in a more consistent and predictable production process, reducing the likelihood of batch failures and maximizing resource utilization. The economic benefits alone, through reduced waste and increased yield, make this a compelling application of biosensor technology.

Biosensors play a crucial role in optimizing food production processes. They can help fine-tune critical parameters such as temperature, pH, and substrate concentration to achieve optimal yields and desired product characteristics. For instance, in dairy processing, biosensors can monitor the activity of enzymes involved in cheese ripening, allowing producers to precisely control the process and achieve consistent quality. Furthermore, biosensors can detect subtle changes in the composition of the product during processing, identifying potential issues before they escalate into major problems. Early detection of off-flavors or undesirable byproducts enables prompt corrective action, preventing product loss and maintaining high quality standards. This ability for proactive intervention translates to substantial cost savings and a significant improvement in overall production efficiency. The advantages extend beyond simply detecting problems; they allow for continuous improvements, driving innovation in food production techniques. By providing real-time feedback, biosensors empower food processors to continuously refine their processes, leading to more efficient and sustainable practices. This data-driven approach to optimization can lead to significant reductions in waste and energy consumption, contributing to both economic and environmental sustainability.

Extending the shelf life of food products is a major challenge, and biosensors are emerging as valuable tools in this area. By continuously monitoring the growth of spoilage microorganisms or the formation of undesirable compounds, biosensors can provide early warnings of impending spoilage. This enables the implementation of preventative measures, such as modifying packaging conditions or adjusting storage temperatures, to significantly prolong the product's shelf life. This translates directly to reduced food waste, a significant problem with major economic and environmental consequences. More importantly, it contributes to improved food security, ensuring the availability of nutritious food for longer periods. Moreover, the accurate prediction of shelf life can facilitate better inventory management, minimizing losses due to spoilage and reducing the need for excessive stock. Beyond microbial growth, biosensors can also monitor the degradation of quality indicators, like texture, color, and aroma, providing a holistic assessment of food quality and shelf life potential. This allows for more precise and effective strategies for preserving food, optimizing storage conditions, and ensuring that consumers receive high-quality products. This ultimately improves consumer satisfaction and reinforces trust in food producers. The ability to accurately predict shelf life also has major implications for labeling and marketing, offering consumers clearer information and increasing product transparency.

5.1 Real-time Process Monitoring

Real-time monitoring within food processing utilizes biosensors to continuously track critical parameters throughout the production line. Imagine a system instantly detecting bacterial contamination in a batch of yogurt, triggering immediate intervention and preventing a costly recall. This isn't science fiction; it's the power of biosensors integrated directly into processing equipment. Such a system provides immediate feedback, optimizing efficiency, and significantly enhancing food safety.

Unlike traditional methods involving time-consuming laboratory analysis, real-time monitoring enables proactive adjustments. For instance, a sudden temperature fluctuation in a fermentation tank can be identified and corrected immediately, preventing spoilage and ensuring product consistency. This reduces waste, saving resources and enhancing profitability for food producers. The data generated provides invaluable insights into the entire processing chain, revealing potential bottlenecks or inefficiencies.

The integration of these biosensors requires careful planning and consideration. The chosen sensors must be robust, reliable, and capable of operating continuously within the harsh environment of a food processing facility. Regular calibration and maintenance are essential to guarantee accurate data. Data management and analysis software play a crucial role, transforming the raw sensor readings into actionable information.

Specific examples of real-time monitoring applications include monitoring pH levels during fermentation, tracking dissolved oxygen in beverage production, or detecting enzyme activity in cheese ripening. By integrating biosensors at multiple stages of the process, a comprehensive overview of product quality and safety is achieved. This level of detail informs process optimization, leading to improved yield, enhanced product quality and reduced production costs.

One critical consideration is the selection of appropriate biosensors. The choice depends on several factors, including the specific parameter being monitored, the characteristics of the food matrix, and the required sensitivity and specificity. Electrochemical biosensors, known for their sensitivity and ease of use, are well-suited for monitoring various parameters, while optical biosensors offer non-invasive monitoring capabilities.

Furthermore, data interpretation and integration with existing process control systems require expertise in both biosensor technology and food processing. Effective communication between sensor systems and production control systems is key for real-time response. Training staff on the use and interpretation of the data generated by these systems is also critical to ensure proper implementation and utilization. A poorly integrated or poorly understood system can lead to errors and even increased risk.

The advantages extend beyond immediate process control. The continuous data stream generated provides a wealth of information for long-term process optimization. By analyzing historical data, producers can identify trends, patterns, and correlations, leading to improved efficiency and reduced variability. This detailed data contributes to improved food safety management systems (FSMS), demonstrating regulatory compliance and enhancing consumer confidence.

Implementing real-time process monitoring requires investment in both hardware and software, as well as training for personnel. However, the long-term benefits significantly outweigh the initial costs. Reduced waste, improved product quality, and enhanced food safety contribute to significant cost savings and increased profitability. Moreover, the ability to rapidly respond to potential problems reduces the risk of costly recalls and reputational damage.

The future of real-time monitoring will likely involve the integration of advanced technologies such as artificial intelligence and machine learning for data analysis and predictive modeling. These technologies will enable more sophisticated process control, leading to even greater efficiency, improved product consistency, and enhanced food safety. As biosensor technology continues to advance, expect further integration into all aspects of food production. This offers the prospect of truly intelligent processing systems, anticipating and proactively addressing potential issues. The result is a more sustainable and robust food production system that meets evolving consumer demands for high-quality and safe food products. This sophisticated approach to food processing represents a significant leap forward in the industry's pursuit of excellence.

5.2 Optimization of Food Production

Precise and timely monitoring is crucial for optimizing food production. Biosensors offer a powerful tool to achieve this, providing real-time data on various parameters throughout the entire production process. Imagine a system that instantly detects subtle changes in temperature, pH, or nutrient levels, allowing for immediate adjustments to prevent spoilage and maintain consistent quality. This proactive approach minimizes waste and maximizes yield, leading to significant economic benefits.

Implementing biosensor-based monitoring systems requires careful planning. First, identify the critical control points in your specific food production process. These are the stages where deviations from optimal conditions can most significantly impact the final product's quality and safety. For example, in dairy production, monitoring milk composition during pasteurization is paramount. Similarly, in baking, tracking fermentation processes is essential for consistent dough quality. Once these points are identified, select appropriate biosensors capable of detecting the relevant parameters.

Selecting the right biosensors involves considering several factors. Sensitivity and specificity are paramount; the sensor must accu-

tion to prevent spoilage and potential health risks. This proactive approach, facilitated by biosensor technology, is far more effective than relying on traditional, often delayed, quality control measures. Furthermore, the choice of packaging materials significantly impacts the shelf life of food items. Biosensors can be integrated into packaging to monitor the internal environment, detecting changes in gas composition, temperature, or the presence of volatile organic compounds that are indicative of spoilage. This data allows for real-time assessment of product freshness and provides valuable insights into the effectiveness of the packaging itself. Such information can lead to the development of more protective and responsive packaging materials, thereby extending the product's usable life and reducing food waste. Imagine a smart label on a yogurt container that changes color when the yogurt begins to spoil, offering a clear visual cue to the consumer. This illustrates the potential of biosensors for direct consumer engagement with food quality.

Innovative preservation techniques, such as modified atmosphere packaging (MAP), benefit greatly from biosensor monitoring. Precise measurements of gas concentrations within the package, provided by biosensors, ensure that the optimal atmosphere is maintained to inhibit microbial growth and extend shelf life. This ability to control the gaseous environment is far superior to traditional methods that rely on less precise and less immediate feedback. Moreover, the real-time data provided by biosensors allows for adjustments to the packaging process itself, leading to continuous optimization and improvement.

Beyond direct monitoring, biosensors can contribute to the improvement of shelf life through assisting in the selection of raw materials. Biosensors can quickly and efficiently assess the quality and freshness of ingredients before they enter the processing chain. By identifying products of substandard quality, biosensors can prevent the use of materials prone to rapid deterioration, thereby directly impacting the shelf life of the final product. This is especially useful in the context of large-scale food processing, where the accurate assessment of large quantities of raw materials is critical.

In conclusion, biosensors offer a versatile and powerful toolkit for extending food shelf life. From monitoring crucial quality parameters during processing, to integrating into smart packaging for real-time environmental analysis, to optimizing preservation techniques and ensuring ingredient quality, biosensors provide a suite of benefits. Their adoption represents a significant step towards a more efficient, sustainable, and safe food production and distribution system, ultimately reducing waste and improving food security for all. The use of biosensors in this context will undoubtedly continue to develop and play an increasingly pivotal role in the future of food technology. The cost-effectiveness and ease of use of biosensors will inevitably improve, making them accessible to a broader range of food producers, small-scale to multinational.

6 Future Trends and Challenges

Miniaturization is revolutionizing biosensor design. We're seeing the development of lab-on-a-chip devices, incredibly small, portable systems capable of performing complex analyses. Imagine a handheld device capable of detecting multiple pathogens or toxins simultaneously, providing near-instantaneous results at the point of need, whether it's a farm, processing plant, or even a consumer's kitchen. This trend promises significant improvements in speed and convenience, making biosensors accessible to a much wider range of users. However, miniaturization also presents challenges; maintaining sensitivity and accuracy at such small scales requires innovative approaches to signal transduction and data processing.

The integration of nanotechnology offers exciting prospects. Nanoparticles, with their unique properties, enhance sensitivity and selectivity in biosensors. For example, gold nanoparticles can amplify the signal from a biological interaction, leading to improved detection limits. Carbon nanotubes offer exceptional electrical conductivity, facilitating faster and more efficient signal transduction. Similarly, quantum dots provide bright, stable fluorescence, useful in optical biosensors. But the use of nanomaterials requires careful consideration of their potential toxicity and environmental impact, necessitating rigorous safety assessments before widespread application.

Standardization remains a significant hurdle. Lack of uniform proto-

cols for biosensor development, validation, and performance evaluation hinders widespread acceptance and interoperability. Establishing internationally recognized standards for biosensor design, manufacturing, and data reporting is crucial. This would not only improve reliability and comparability of results but also facilitate the regulatory approval process, paving the way for greater market penetration. Collaboration between research institutions, industry, and regulatory bodies is vital in achieving this goal. Without a coordinated effort, the field risks fragmentation and limitations in widespread adoption.
Cost-effectiveness is paramount for widespread adoption, particularly in developing countries where food safety issues are often most acute. The current cost of many biosensors can be prohibitive, limiting their accessibility to smaller producers or laboratories. Reducing manufacturing costs through innovative materials, streamlined production processes, and economies of scale are essential. Exploring alternative manufacturing techniques, such as 3D printing, could also significantly reduce costs. Furthermore, developing simpler, less expensive biosensors specifically tailored for low-resource settings is critical for maximizing the global impact of this technology.
Regulatory approval is crucial, yet often challenging. Biosensors, like any analytical technology used for food safety, must meet rigorous regulatory requirements to ensure accuracy, reliability, and safety. The process of obtaining regulatory approval can be lengthy and expensive, hindering the timely introduction of new biosensor technologies to the market. Streamlining the regulatory pathway, through clear guidelines and expedited review processes, is essential to encourage innovation and accelerate the adoption of safe and effective biosensors. Clear communication and collaboration between developers and regulatory bodies are critical to navigate this complex landscape.
Ethical considerations need careful consideration. Biosensors can generate vast amounts of data related to food safety and quality. Safeguarding the privacy and security of this data is paramount. Ethical guidelines and data protection policies must be established to ensure responsible data handling and prevent misuse. Transparency

and accountability in data management are crucial to build public trust and foster the responsible use of this technology. Further considerations involve the potential displacement of labor due to automation, prompting a need to plan for workforce retraining and upskilling opportunities. Addressing these ethical concerns proactively is essential to ensure the beneficial application of biosensor technologies. Ignoring the ethical dimensions risks public resistance and hampered development.

Commercialization presents both opportunities and challenges. Successfully transitioning biosensor technologies from the laboratory to the marketplace requires careful planning and execution. This includes identifying target markets, developing effective marketing strategies, securing funding, and establishing reliable supply chains. Collaboration between academic researchers, industry partners, and investors is crucial for successful commercialization, facilitating knowledge transfer, technology development, and market penetration. Addressing technical challenges, obtaining necessary regulatory approvals, and securing market share are all crucial components that require strategic planning and collaboration.

6.1 Advancements in Biosensor Technology

Miniaturization is revolutionizing biosensor design. Smaller devices mean greater portability, enabling on-site testing in diverse environments, from processing plants to remote farms. This eliminates the need for sample transport, reducing both time and potential contamination issues. Furthermore, microfluidic systems, integrated with miniaturized biosensors, are automating sample preparation and analysis, reducing hands-on time and increasing throughput. Imagine a handheld device capable of rapidly identifying a specific pathogen in a food sample with minimal training needed to operate it. That's the power of miniaturization. The result is a more efficient and cost-effective method for food safety assessment.

The integration of nanotechnology offers remarkable opportunities to enhance biosensor performance. Nanomaterials, such as nanoparticles and nanotubes, possess unique physicochemical properties that significantly improve sensitivity, selectivity, and stability. For example, gold nanoparticles are frequently used to amplify signals in electrochemical biosensors, leading to faster and more accurate detection of foodborne pathogens. Carbon nanotubes exhibit exceptional electrical conductivity, making them ideal components for highly sensitive biosensors. This combination of enhanced characteristics promises faster results, improved reliability, and the ability to detect even trace amounts of contaminants.

Lab-on-a-chip (LOC) technology is emerging as a highly promising approach, merging microfluidics and biosensing into a single platform. These miniaturized laboratories can perform multiple analytical steps, including sample preparation, separation, detection, and analysis, all on a single chip. This integrated approach streamlines the entire detection process, decreasing analysis time and improving the accuracy of results. Furthermore, LOC devices offer improved portability, enabling point-of-care testing in resource-limited settings. The potential of these devices to revolutionize food safety monitoring, particularly in remote or developing regions, is immense.

Impedimetric biosensors, which measure changes in electrical impedance as a result of biomolecular interactions, are gaining significant attention. These sensors are known for their simplicity, cost-effectiveness, and real-time monitoring capabilities. Their ability to provide continuous data during food processing is invaluable for identifying potential contamination early on. Imagine a system continuously monitoring a production line for the presence of harmful bacteria, alerting personnel immediately if contamination occurs, allowing for rapid response and minimized losses. Impedimetric technology is transforming real-time quality control in food processing.

Optical biosensors, leveraging techniques like surface plasmon resonance (SPR) and fluorescence, offer exceptional sensitivity and multiplexing capabilities. SPR biosensors, for instance, can detect subtle changes in refractive index, providing highly sensitive and label-free detection. Fluorescence-based biosensors are particularly adept at detecting specific molecules through the use of fluorescent probes.

Multiplexing capabilities allow simultaneous detection of various foodborne pathogens or toxins, maximizing efficiency and reducing analysis time. These advanced technologies pave the way for sophisticated food safety control mechanisms.
Wireless and remote sensing are adding another layer of sophistication. Wireless biosensors allow for continuous monitoring of food products without requiring direct physical access. This has implications for long-term storage facilities, transportation monitoring, and smart packaging solutions. These systems can transmit data wirelessly to a central monitoring system, providing real-time insights into food quality and safety parameters. Combining this technology with machine learning algorithms will eventually create early warning systems capable of proactively addressing potential threats to food safety.
The development of biosensors tailored to specific food matrices is crucial. The complexity of food composition often interferes with sensor performance. Biosensors optimized for specific food types, like dairy products or meat, will yield more accurate and reliable results. Addressing this matrix effect will improve the overall effectiveness of biosensor-based food safety assessments. This approach accounts for the complex nature of food materials and yields more precise analytical data.
Despite these advancements, challenges persist. The high cost of certain biosensors remains a barrier to widespread adoption, particularly for smaller food producers. Therefore, developing cost-effective manufacturing processes and reusable sensor platforms is crucial for broader implementation. Furthermore, standardization and regulatory approval are vital for ensuring the reliability and acceptance of biosensor technologies within the food industry. Consistent and reliable outcomes for a variety of applications need to be guaranteed.

6.2 Challenges and Limitations

Despite the significant potential of biosensors in revolutionizing food safety and quality control, several challenges and limitations hinder their widespread adoption. Cost remains a major barrier, particularly for smaller food producers who may lack the resources to invest in sophisticated biosensor technology. The initial investment in equipment, specialized reagents, and skilled personnel can be substantial, creating a significant hurdle for implementation. Moreover, the ongoing maintenance and calibration of these systems add to the overall expenditure, potentially making them cost-prohibitive for some operations. This financial constraint directly impacts the accessibility and feasibility of biosensor integration, especially within developing countries.

Many currently available biosensors lack the necessary robustness and stability for routine use in harsh food processing environments. Some biosensors are highly sensitive to temperature fluctuations, variations in pH, or the presence of interfering substances within food matrices. These sensitivities can lead to inaccurate readings, compromised reliability, and ultimately, questionable results. Consequently, the need for careful control of environmental conditions and meticulous sample preparation often adds complexity and expense, further restricting applicability. The development of more rugged and resilient biosensors is critical for broader acceptance within the industry.

Standardization and validation of biosensor technologies are crucial for ensuring data comparability and regulatory compliance. Currently, a lack of universally accepted standards and protocols makes it difficult to compare results obtained from different biosensor systems. This inconsistency challenges the reliability of data interpretation and hinders the ability to establish universally recognized benchmarks for food safety and quality. International collaborations and collaborative efforts towards developing standardized methodologies and reference materials are urgently needed to address this limitation. Furthermore, achieving regulatory approval for new biosensors can be a protracted and complex process, which presents a significant obstacle for commercialization.

The specificity and sensitivity of biosensor systems vary greatly depending on the target analyte and the chosen biosensor type. While some biosensors can detect specific pathogens or toxins with remarkable precision, others may exhibit cross-reactivity with similar compounds, leading to false-positive or false-negative results. This potential for inaccuracy raises concerns about the reliability of biosensor data and the impact on decision-making related to food safety and quality control. Therefore, meticulous selection of appropriate biosensor technologies and rigorous validation are essential to minimize errors and ensure the accuracy of results. Improving the sensitivity and specificity of existing biosensors through advancements in materials science and biomolecular engineering remains a critical area of research.

Another significant challenge lies in the limited availability of user-friendly and portable biosensor devices suitable for on-site testing. Many currently available biosensors require sophisticated laboratory equipment and trained personnel, thus limiting their applicability for rapid, point-of-care testing in food processing plants or during transportation. The development of miniaturized, portable, and user-friendly biosensors would significantly enhance their practical utility, enabling faster responses to potential food safety risks and more efficient quality control measures. This requires innovations in sensor design, signal processing, and power management to create robust yet portable devices capable of performing accurately and reliably outside a laboratory environment. Such developments are crucial for expanding the potential reach and impact of biosensors within the food industry.

Finally, concerns about the long-term stability and shelf life of biosensor components also hinder their widespread implementation. Some biological recognition elements, such as antibodies or enzymes, can lose their activity or undergo degradation over time, leading to a decrease in sensor performance. Maintaining the integrity and functionality of these components during storage and use is crucial for ensuring the reliability and accuracy of the biosensor system. Further research into developing more stable biological recognition elements and improving the design of biosensor devices to protect these elements from environmental influences are vital steps towards increasing the longevity and practical utility of these technologies. Addressing these challenges through collaborative research, technological innovation, and regulatory harmonization is crucial to unlock the full potential of biosensors in safeguarding food safety and enhancing food quality.

6.3 Regulatory Considerations

Navigating the regulatory landscape for biosensor implementation in food safety and quality control can feel like traversing a complex maze. Success hinges not just on technological prowess but also on a thorough understanding and meticulous adherence to existing regulations. These regulations, which vary significantly across jurisdictions, dictate everything from the design and validation of your biosensor to its manufacturing, distribution, and ultimately, its use in food testing. Ignoring these aspects can lead to costly delays, product recalls, and even legal repercussions. Therefore, proactive planning and consultation with relevant authorities are paramount.
Understanding specific regulations requires diligent research. For example, the Food and Drug Administration (FDA) in the United States has specific guidelines for analytical methods used in food testing, including those employing biosensors. These guidelines often emphasize accuracy, precision, and the validation of the method's performance characteristics. Similarly, in the European Union, regulations like those under the General Food Law Regulation (EC) No 178/2002, and the specific regulations for individual food categories, set rigorous standards for food safety and the methods used to ensure compliance. It's crucial to identify the specific regulations governing your target market.
Beyond overarching food safety regulations, specific biosensor types may fall under additional regulatory scrutiny. For instance, biosensors using genetically modified organisms (GMOs) might require compliance with regulations governing the use of GMOs in food production and testing. Similarly, biosensors incorporating nanomaterials may necessitate adherence to evolving guidelines regarding the

safety and environmental impact of nanotechnology. Always research the regulations specifically related to any novel materials or technologies used in your biosensor.

Furthermore, the process of obtaining regulatory approval for a new biosensor often involves a rigorous validation process. This process demonstrates the accuracy, precision, and reliability of the biosensor under various conditions and demonstrates its suitability for its intended purpose. Thorough documentation of validation studies, including detailed experimental protocols, data analysis, and quality control measures, is crucial to supporting your application for regulatory approval. This process is often time-consuming and resource-intensive, but it is essential for gaining market access.

The regulatory landscape extends beyond the biosensor itself to include the data generated by its use. Data integrity and traceability are paramount. A robust quality management system (QMS) is crucial, ensuring that data collected is accurate, reliable, and can be readily traced back to the origin. This often involves complying with standards such as ISO 9001 for quality management, ISO 17025 for testing laboratories, or other industry-specific standards relevant to your biosensor's application. Failure to maintain impeccable data records can compromise the validity of your test results and jeopardise your compliance with regulations.

Beyond formal regulations, consider the importance of ethical considerations. Transparency and responsible innovation are key. Clearly communicate the limitations of your biosensor, ensuring the results are interpreted appropriately and avoiding potential misinterpretations. Engage with stakeholders, including industry peers, regulatory bodies, and consumer groups, to promote understanding and build trust in your technology. Proactive engagement can help to shape future regulations in a beneficial way. This holistic approach, encompassing both legal compliance and ethical responsibility, is vital for the successful integration of biosensors into the food safety system. By understanding and proactively addressing these regulatory hurdles, you can pave the way for widespread adoption of biosensor technology, contributing significantly to enhancing food safety and quality control globally.

6.4 Commercialization Aspects

Successful commercialization of biosensor technology for food safety hinges on several key factors. Firstly, a strong understanding of the target market is crucial. Are you focusing on large-scale food processors needing high-throughput systems, or smaller businesses needing rapid, on-site testing? Defining your ideal customer allows you to tailor your product development and marketing strategies effectively. This precise targeting will significantly impact your overall success.
Next, consider the regulatory landscape. Navigating the complex web of food safety regulations—varying between countries and regions—requires careful planning and potentially significant investment. Compliance with standards like those set by the FDA or equivalent agencies is paramount for market entry and maintaining credibility. This process can be lengthy and demanding, requiring expert legal and regulatory advice.
Cost-effectiveness is another pivotal aspect. While biosensors offer significant advantages, their price must be competitive to attract widespread adoption. Minimizing manufacturing costs without compromising performance is key. This often involves exploring different manufacturing techniques and material sourcing strategies. This is a crucial element that many companies overlook.
User-friendliness is often overlooked, but vitally important. The technology, however sophisticated, must be accessible to the intended users, whether highly trained technicians in a laboratory or less experienced personnel in a processing plant. Intuitive design, simplified operation, and minimal training requirements will significantly enhance adoption rates. This must be carefully considered from the very beginning of the design process.
Marketing and distribution strategies are vital in reaching the target market. This includes developing a clear and concise message that highlights the unique value proposition of the biosensor. It may involve showcasing its speed, accuracy, and cost-effectiveness com-

pared to existing methods. Building strong relationships with key distributors and forming strategic alliances can open doors to new markets and increase market penetration.

Furthermore, intellectual property protection is essential. Securing patents or other forms of intellectual property rights will safeguard your investment and provide a competitive edge in the market. A robust IP strategy can prevent competitors from replicating your technology and allow you to capitalize on your innovation. This is a critical step which is often underestimated.

Finally, consider long-term sustainability. Building a reputation for reliability and providing excellent customer support will foster trust and loyalty. Continuous improvement of the technology and adaptation to emerging needs will ensure its relevance in a constantly evolving food safety landscape. Investing in research and development is a vital step to improve the longevity of your product in the market. This long-term vision is essential for achieving sustainable success.

The commercial success of biosensors in the food industry depends on a carefully orchestrated approach. It's not just about technological innovation; it is about understanding the market, navigating regulations, ensuring cost-effectiveness, and building a strong brand. By focusing on these key aspects, researchers and entrepreneurs can unlock the significant potential of biosensors for improving food safety and quality globally. Strategic partnerships, robust marketing, and a forward-looking approach will pave the way for sustained growth and a positive impact on the global food supply chain. It requires a commitment not just to the technology itself, but to building a successful and enduring business around it. And this business requires careful planning, meticulous execution, and a willingness to adapt to the ever-changing demands of the market. Without this holistic approach, even the most innovative technology may fail to reach its full potential.

6.5 Ethical Implications

The widespread adoption of biosensors in food safety and quality control presents a unique ethical landscape. We must consider the potential for misuse, the impact on consumer trust, and the equitable distribution of this technology. The accuracy and reliability of biosensor data are paramount; inaccurate results could lead to unnecessary recalls or, worse, the release of contaminated food. This necessitates rigorous validation and quality control measures across the entire biosensor lifecycle, from manufacturing to data interpretation. Transparency in reporting methods and data provenance is crucial for building public confidence. Any biases inherent in the design or application of the technology must be identified and addressed to prevent discriminatory outcomes. For instance, ensuring that biosensor-based food safety systems are equally accessible to small-scale producers is crucial for supporting fair competition and preventing the disproportionate impact on smaller businesses. These considerations move beyond purely technical specifications, requiring a nuanced approach to the integration of this technology.

Data privacy is another critical ethical concern. Biosensors generate data, and the collection, storage, and use of this data must adhere to strict ethical guidelines. Protecting the privacy of consumers and businesses involved in the food supply chain is non-negotiable. Clear protocols regarding data security, informed consent, and data ownership should be established and strictly enforced to prevent misuse or unauthorized access. The potential for data manipulation or unauthorized use of sensitive food safety information for commercial or malicious purposes necessitates a robust and transparent data governance framework. This framework should incorporate mechanisms for data anonymization, encryption, and audit trails to maintain accountability and prevent any breaches of trust. Clear regulations regarding data ownership and access rights will safeguard the integrity of the system and prevent any abuse of this sensitive information.

The potential for job displacement due to automation driven by biosensors must also be acknowledged. As biosensors automate various tasks in food production and quality control, jobs may be affected. Mitigation strategies need to be considered, such as retraining and reskilling programs to equip workers with the necessary skills for new roles in the evolving food industry. This is not just an economic consideration but also an ethical responsibility to prevent social disruption and ensure a just transition for the affected workforce. We have a responsibility to minimize the negative impacts on human employment while maximizing the benefits of technological advancements. This requires a proactive approach to workforce development, ensuring that skills training initiatives align with the changing needs of the food industry.

Beyond the direct impacts, the broader ethical implications of biosensor technology on society must be addressed. The increasing reliance on technology for food safety decisions raises questions about the role of human judgment and expertise. Striking a balance between automated systems and human oversight is crucial. Over-reliance on biosensors without sufficient human review could lead to errors and oversight. We must ensure that human expertise continues to play a vital role in interpreting sensor data and making informed decisions regarding food safety and quality. This integrated approach allows the benefits of technology to enhance human capabilities and maintain a level of human oversight to prevent biased outcomes. We need to develop effective communication strategies to bridge the gap between technological advancements and public understanding, promoting trust in biosensor technologies while emphasizing the ongoing importance of human expertise.

Finally, the equitable access to biosensor technology is a crucial ethical consideration. Biosensor technologies must be accessible to all stakeholders in the food supply chain, from large-scale producers to small farms and individual consumers. This implies a focus on affordable and user-friendly biosensor devices, coupled with training and educational resources to empower users. Any disparities in access could exacerbate existing inequalities, potentially widening the gap between developed and developing nations. The goal should be to promote the widespread adoption of biosensors to improve food safety and quality across the globe. This necessitates international collaboration to address the challenges of affordability, accessibility, and capacity building in low-resource settings. A collaborative, globally inclusive approach to the development and implementation of biosensor technologies is essential for maximizing their benefits for society.

7 Conclusion and Perspectives

This book has explored the transformative potential of biosensors across the food industry spectrum. From identifying harmful pathogens and toxins to monitoring quality parameters and optimizing processing, biosensors offer unprecedented speed, accuracy, and efficiency. We've seen how diverse biosensor types, each with its strengths and weaknesses, contribute to a comprehensive safety and quality control system. Electrochemical, optical, and piezoelectric approaches, coupled with techniques like ELISA and PCR, have proven their efficacy in detecting a range of contaminants, ensuring both consumer and regulatory compliance. The ability to quickly analyze food products for bacteria, viruses, parasites, mycotoxins, pesticides, and heavy metals is paramount in preventing outbreaks and ensuring safe consumption.

The application of biosensors extends beyond detection to the optimization of food production. Real-time monitoring of crucial parameters allows for immediate adjustments, minimizing waste and maximizing yield. By integrating biosensors into the production line, producers can enhance efficiency and ensure consistent quality, reducing the risk of spoilage and ensuring prolonged shelf life. This translates into both economic benefits for businesses and enhanced food security for consumers. This direct and immediate feedback loop is a powerful tool in modern food production.

However, realizing the full potential of biosensor technology requires addressing several key challenges. Cost-effectiveness remains a hurdle, particularly for smaller food producers. Standardization across different biosensor platforms is also crucial for consistent and reliable results. Regulatory frameworks need to adapt and evolve to accommodate the rapid pace of biosensor development, ensuring that new technologies are both effective and safe. This interplay between technological advancement and regulatory oversight will determine widespread adoption.

Future research should focus on several promising avenues. Miniaturization and portability are key objectives, leading to more user-friendly and readily deployable devices suitable for both laboratory and on-site analyses. Integrating biosensors into intelligent packaging systems offers the potential for real-time monitoring throughout the supply chain, from farm to table, providing a complete picture of food safety and quality. The development of more robust and cost-effective biosensors, specifically tailored to detect emerging threats and contaminants, is also critical.

Beyond technological improvements, research should also address the ethical implications of widespread biosensor adoption. Data security and privacy considerations, alongside the potential for misuse or manipulation of biosensor data, need careful consideration. Public awareness and education on the benefits and limitations of biosensor technology are equally essential. Addressing these ethical considerations will build trust and ensure responsible deployment of this powerful technology.

In conclusion, the integration of biosensors into food safety and quality control represents a significant advancement, offering a powerful toolset for safeguarding food supplies and ensuring consumer health. While challenges remain, the opportunities presented by continued research and development are immense, pointing towards a future where biosensors are integral to a safe, efficient, and sustainable food system. The collaborative efforts of scientists, engineers, regulators, and industry stakeholders will determine the extent to which this technology transforms the global food landscape. The benefits, however, clearly outweigh the challenges, promising a future of enhanced food security for all. The path forward requires a commitment to innovation, ethical considerations and effective regulatory frameworks, resulting in more reliable, accessible, and affordable biosensor technologies for the benefit of everyone. The widespread adoption of this technology would represent a profound shift in food production and safety. This shift offers great potential for enhancing global food security and public health.

7.1 Summary of Findings

Our exploration of biosensors reveals a transformative potential across the food industry. From the initial chapters establishing fundamental principles and categorizing sensor types, to the detailed application sections covering pathogen, toxin, and quality parameter detection, a clear picture emerges: biosensors offer a powerful suite of tools for enhancing food safety and quality. The rapid, sensitive, and often on-site detection capabilities dramatically improve response times to contamination events, minimizing risks to public health and economic losses.

The efficacy of biosensors in detecting various foodborne pathogens—bacteria, viruses, and parasites—has been consistently demonstrated throughout this work. Different methodologies, ranging from traditional ELISA adaptations to advanced PCR integrations, provide a tailored approach based on the specific threat. This versatility, coupled with the increasing miniaturization and portability of sensor technology, promises broader accessibility, even in resource-limited settings.

Similarly, the detection of food toxins—mycotoxins, pesticides, heavy metals—presents a significant challenge addressed with notable success using biosensor technologies. The high sensitivity and specificity achieved allow for precise quantification, vital for regulatory compliance and consumer protection. This ability to identify and measure even trace amounts of harmful substances underscores the crucial role of biosensors in safeguarding food supplies.

Moving beyond hazard detection, biosensors also prove invaluable in monitoring food quality parameters. Assessing freshness through the detection of spoilage indicators and analyzing nutritional content provide consumers with crucial information about product quality, ultimately driving informed purchasing decisions and reducing food waste. The non-destructive nature of many biosensor methods further strengthens their appeal for applications across the entire supply chain.

The integration of biosensors into food processing workflows offers transformative potential for real-time monitoring, optimization, and shelf-life extension. Continuous monitoring of parameters like temperature, pH, and microbial load allows for immediate adjustments, improving efficiency and reducing waste. Furthermore, the optimization of food production through precise control of environmental factors and processing parameters ultimately contributes to greater yields and improved product quality.

While significant progress has been made, the full realization of biosensor technology's potential hinges on addressing several key challenges. The ongoing need for cost reduction and standardization ensures widespread accessibility and adoption. Navigating regulatory hurdles and addressing ethical considerations regarding data privacy and consumer confidence are paramount for long-term success. Further research into user-friendly and portable devices is necessary to fully unlock the practical advantages of this technology.

However, despite these challenges, the future looks bright for biosensors in the food industry. Continuous advancements in miniaturization, sensitivity, and integration with other technologies promise even greater capabilities. The potential for widespread application extends beyond the examples detailed in this book, spanning diverse areas of food production, processing, and distribution. The development of more sophisticated algorithms and machine learning integration will further enhance the speed, accuracy, and interpretation of sensor data, fostering real-time decision-making capabilities across the industry. The combination of advancements in nanotechnology and microfluidics opens avenues for the development of highly sensitive, portable, and low-cost devices, making biosensors accessible to a wider range of users, from small-scale producers to large-scale industrial operations. The development of innovative biosensors using advanced materials and principles, such as aptamers and molecularly imprinted polymers, may also improve the sensitivity, selectivity, and stability of existing technologies. The commercial potential is immense, and successful implementation will have a profound and lasting impact on food security, public health, and economic sustainability. This technology represents a key component in achieving a more resilient and efficient food system for the future, ultimately benefiting both producers and consumers alike. The convergence of various scientific disciplines—biology, chemistry, engineering, and information technology—continues to fuel progress in biosensor development, paving the way for a new era of food safety and quality control. Future research directions should also focus on developing standardized protocols and guidelines for biosensor application and data interpretation to ensure consistency and reliability across various settings. This standardization is crucial for achieving widespread acceptance and facilitating regulatory approval. The synergy between biosensors and other advanced technologies, such as blockchain and Internet of Things (IoT) systems, holds the promise of building a more robust and secure food supply chain. These integrated systems can enhance traceability, transparency, and accountability, providing consumers with increased confidence in the safety and quality of food products. As the technology continues to mature and become more accessible, the potential for even greater transformative changes across the food sector is undeniable.

7.2 Future Research Directions

The immediate future of biosensor technology in food safety hinges on addressing limitations in current methodologies. While impressive advancements have been made, the need for more robust, portable, and user-friendly devices remains paramount. Imagine a world where rapid, on-site testing for pathogens and toxins is as commonplace as using a home pregnancy test – this requires miniaturization and simplification of current bulky and lab-based systems. Research should focus on developing low-cost, disposable biosensors suitable for use by non-experts, expanding accessibility and potentially transforming how food safety is managed globally.

Developing biosensors capable of detecting multiple food contaminants simultaneously is another crucial area. Current methods often target single contaminants, requiring multiple tests for a comprehensive analysis. A multi-analyte sensor would streamline the process, providing faster, more efficient, and cost-effective screening, leading to quicker responses to contamination events. This requires innovative approaches in sensor design and signal processing to avoid cross-reactivity and maintain accuracy. This can significantly improve the speed and efficiency of food safety checks.

Beyond multiplexing, the integration of artificial intelligence and machine learning in biosensor data analysis presents a significant opportunity. These tools can enhance sensitivity, improve data interpretation, and potentially predict future contamination risks based on historical data. Sophisticated algorithms could analyze complex datasets, recognizing subtle patterns that might be missed by human analysts, leading to more accurate assessments and improved risk management strategies. However, rigorous validation and standardization of these AI-driven tools are essential for widespread acceptance.

The development of novel biorecognition elements is critical. Improving the specificity and affinity of antibodies, aptamers, or enzymes used in biosensors is essential for enhancing their sensitivity and selectivity. This research can explore new molecular recognition technologies, such as engineered proteins with improved binding characteristics or synthetic receptors that mimic biological interactions. Exploration of the potential of CRISPR-Cas systems as detection tools requires additional focus. The potential benefits could greatly influence both testing speed and sensitivity.

Beyond technical advancements, standardization and regulatory frameworks for biosensor technology are equally crucial. Harmonized protocols for sensor validation and performance testing are needed to ensure reliable and comparable results across different laboratories and manufacturers. Clear guidelines regarding the regulatory approval process for new biosensors are essential to expedite their adoption in the food industry. Collaborative efforts between scientists, regulatory bodies, and industry stakeholders are vital in establishing a robust and transparent regulatory landscape.

Finally, a deeper understanding of the economic and societal impacts of widespread biosensor implementation is necessary. Studies focusing on cost-benefit analyses, the economic feasibility of different biosensor technologies, and their impact on various stakeholders (e.g., farmers, food processors, consumers, and policymakers) are needed to facilitate informed decision-making. Addressing ethical considerations regarding data privacy, the potential displacement of workers, and the responsible use of advanced technologies is crucial for ensuring the equitable and sustainable integration of biosensors into the food system. This involves rigorous cost-benefit analysis to justify wide scale adoption. This future research must focus on real-world impact, creating a holistic understanding of the long-term consequences.

www.ingramcontent.com/pod-product-compliance
Ingram Content Group UK Ltd.
Pitfield, Milton Keynes, MK11 3LW, UK
UKHW020155290926
15471UKWH00005B/887

9 798349 389825